1 The variety of life

Wildlife!... what wildlife?

This country is rich in **wildlife**. The drawing shows just a few of the animals and plants that you can see in the countryside.

Q1 Make a list of all the plants you can see in the drawing.

Q2 Make a list of all the animals you can see in the drawing.

Q3 Sort your list of animals into smaller groups. Put the animals into groups which have some **features** in common.

Group 1	Group 2	Group 3	Group 4	Group
hawk	mouse			

Q4 Describe what features the animals in each group have in common.

Simple plants

On a walk round the school grounds, you will see many different sorts of plants. All these plants can be put into one of two groups. They are either simple **non-flowering** plants or **flowering** plants.

▶ Simple plants such as **mosses** and **ferns** do not have flowers and seeds. They reproduce using **spores**. These spores are carried away by the wind. If they land in a suitable place, a new plant will start to grow.

Apparatus

- ☐ clipboard ☐ pencil
- ☐ map of school and grounds

Protect living things. Do not damage the plants.

A plant survey

In this survey you are going to find out where mosses and ferns like to grow.

A Look around the school grounds for mosses and ferns. Look carefully; they are not always easy to find. ▲

B Show on your map where you found these simple plants. ▲

Use symbols such as:
mosses – M
ferns – F

Q1 Where did you find mosses growing?

Q2 Where did you find ferns growing?

Q3 In what areas do simple plants like to grow?

Q4 My neighbour wants to plant a large fern plant in the middle of a flowerbed. Do you think this is a good spot for the fern? Explain your answer.

Extension exercise 1 can be used now.

Flowering plants

Many of the plants growing in the school grounds belong to the group called 'the flowering plants'. Flowers produce seeds by joining **pollen grains** from one flower with **ovules** from another. The process of moving the pollen to a second flower is called **pollination.**

▲ Plants like daisies and buttercups have brightly coloured flowers, sweet scents and sugary **nectar** to attract insects. As an insect feeds inside the flower, it becomes covered in pollen grains. When the insect moves to the next flower, some of this pollen is brushed off. This is called **insect-pollination**.

▲ Plants like grasses use the wind for pollination. Their flowers are usually small and green because they do not need to attract insects. Each flower can produce hundreds of thousands of pollen grains which are scattered into the wind. By chance, some pollen grains may land on another flower and pollinate it.

▶ **Hayfever** is caused by pollen grains in the the air. Some people are **allergic** to the pollen and suffer from runny noses, itchy eyes and wheezy chests. The pollen count shows how much pollen is floating in the air. It can be very high in June and July as the grasses release their pollen in these months.

Q1 What is pollination?

Q2 Name one plant which is wind-pollinated.

Q3 Name four plants which are insect-pollinated.

Q4 What is hayfever?

Q5 The table below shows how the pollen count has varied over a period of six weeks. When would hayfever sufferers be most uncomfortable?

Date	Pollen count (grains per cubic metre)
15th May	45
22nd May	40
29th May	55
5th June	110
12th June	180
19th June	120

Identifying common flowering plants

If you look carefully at the plants growing in the school grounds, you will notice that there are many different sorts (**species**). In this survey you are going to try to find the common plants drawn below. ▼

Apparatus

- ☐ map of school and grounds
- ☐ clipboard

Protect living things. Do not pick the flowers.

A Choose an area in the school grounds. It can be on the fields, near a hedge or under some trees. Look carefully around the area for any of the plants. ▲

B Record on your map any plants you find. Use symbols to help you record your results. Repeat the search in as many areas as possible. ▲

Q1 How many of the plants did you find?

Q2 Which plant was the most common?

Q3 Which area had the greatest variety of plants?

Q4 A group of students noticed that some plants grew well near paths and playgrounds even though they were often trodden on. Other plants never grew in these areas. From your results, which plants can survive being trodden on?

Daisies

Daisies grow well in school grounds. In this survey, you are going to find out exactly where they grow best.

Apparatus

- ☐ map of school and grounds
- ☐ clipboard and paper
- ☐ 1 metre square **quadrat** (or four 1 metre rules)

Protect living things. Do not pick the flowers.

Q1 Copy this table.

Area	Number of daisies in $1 m^2$			Average number of daisies in $1 m^2$
	1st	2nd	3rd	
football field				

A Choose any area in the school grounds. It can be on the field, under trees, or near a hedge. Put your quadrat down anywhere (**randomly**) in this area. Try not to choose a particular spot. ▲

B Count the number of daisies inside the quadrat. Write down your results in the table. Put the quadrat down in two other parts of this area. Complete your table. ▲

C Repeat **A** and **B** for as many areas as possible. Complete the table for each area.

Q2 In which area did you find most daisies?

Q3 Why did you put the quadrat down randomly?

Q4 Sally found 50 daisies in a 1 metre square of the school field. How many daisies are there altogether in this 800 square metre field?

Q5 How can you make Sally's estimate of the number of daisies in the field as accurate as possible?

Q6 Jonathon thinks that you get more daisies in grass which is cut more often than in grass that is left to grow. How can you prove whether he is right or wrong?

All animals can be put into one of two groups:

1 **Invertebrates** – animals without a backbone (*e.g.* worms, flies)
2 **Vertebrates** – animals with a backbone (*e.g.* humans, cats)

Sorting out Invertebrates

The large group of invertebrates can be divided up into smaller sets. Each set contains animals with some features in common.

Invertebrates

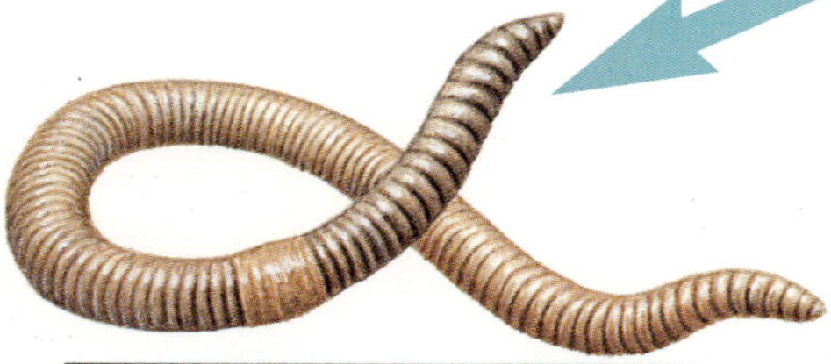

Annelids
- ☐ long **segmented** body
- ☐ no **jointed** legs

Arthropods
- ☐ segmented body
- ☐ jointed **exoskeleton**
- ☐ several pairs of legs
- ☐ **compound eye**
- ☐ **antennae**

Molluscs
- ☐ soft body
- ☐ single slimy foot
- ☐ some with shell

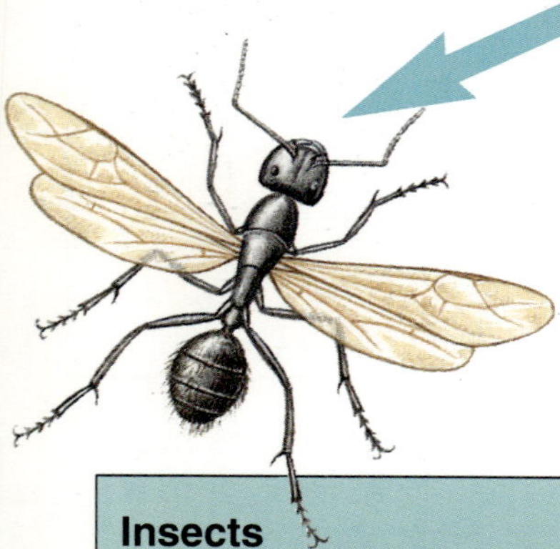

Insects
- ☐ three pairs of legs
- ☐ wings
- ☐ three parts to body

Arachnids
- ☐ four pairs of legs
- ☐ two parts to body

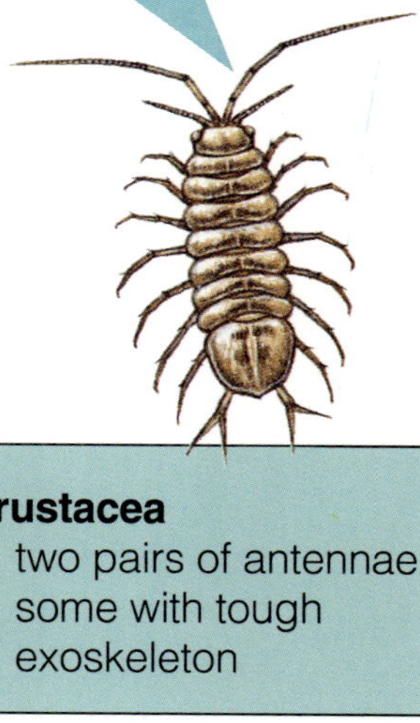

Crustacea
- ☐ two pairs of antennae
- ☐ some with tough exoskeleton

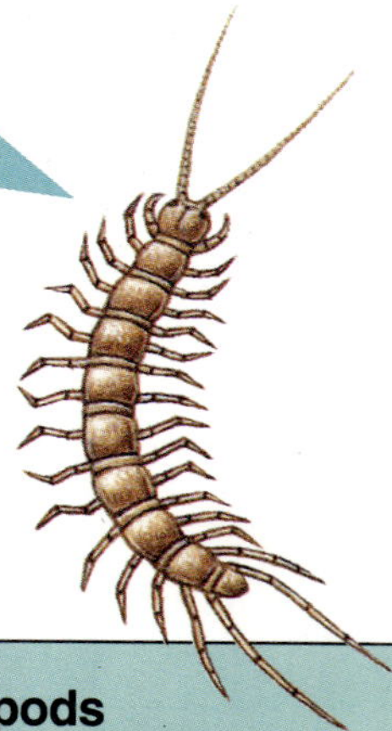

Myriapods
- ☐ long segmented body
- ☐ jointed legs on each segment

Q1 How many legs do insects have?

Q2 Name a common arachnid.

Q3 How many legs do arachnids have?

Q4 To which group do earthworms belong?

Q5 State one difference between an earthworm and a millipede.

Q6 Emma, a farmer's daughter, found this creature in the wool of a sheep. To which set does it belong?

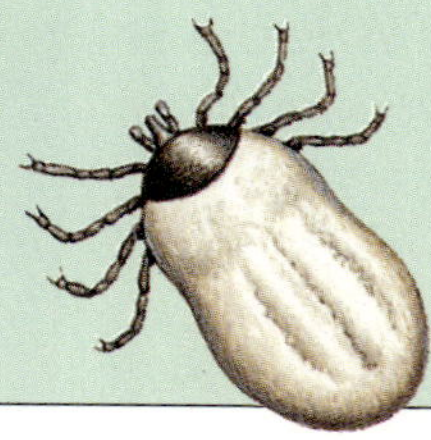

Looking at leaf litter

Leaf litter is the layer of dead and rotting leaves left on the top of the ground when trees lose their leaves in autumn. It is home for many invertebrates. Some of these animals feed on the leaves, breaking them down and making them rot. These animals are called **decomposers** and are needed to get rid of dead plants and animals. Without them we would be buried under great piles of dead leaves.

In this experiment, you will find and name some of the invertebrates living in leaf litter.

Apparatus

- ☐ leaf litter ☐ newspaper
- ☐ tweezers ☐ magnifying lens
- ☐ sample tubes ☐ bench lamp
- ☐ small soft paintbrush

Protect living things. Handle the invertebrates with care.

A Put a handful of leaf litter onto the newspaper. Shine the bench lamp onto the leaf litter. ▲

B Look *carefully* through the leaf litter. Gently brush any animals you find into sample tubes. Collect as many different species as you can. ▲

Wash your hands thoroughly with soap and water after handling leaf litter.

C Use the information and drawings on pages 8 and 9 to name the invertebrates you have found.

Q1 List the invertebrates you found.

Q2 Draw one of your invertebrates.

Q3 Which invertebrate was the most common?

Q4 Name two invertebrates which feed on plants.

Q5 Name two invertebrates which feed on other animals.

Q6 Why are decomposers so important?

Identifying invertebrates

The drawings on pages 8 and 9 will help you to identify the invertebrates in your leaf litter. You may not find all of them, or you may find more than those shown. If in doubt, ask your teacher for advice.

1 *Annelids*

Earthworm
(Total length 5–20 cm)

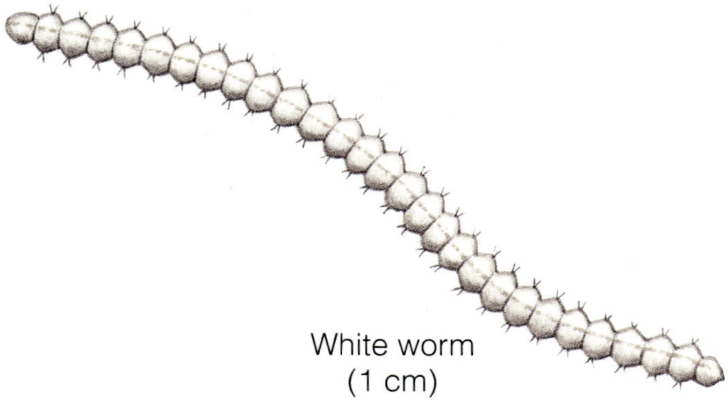

White worm
(1 cm)

2 *Molluscs*

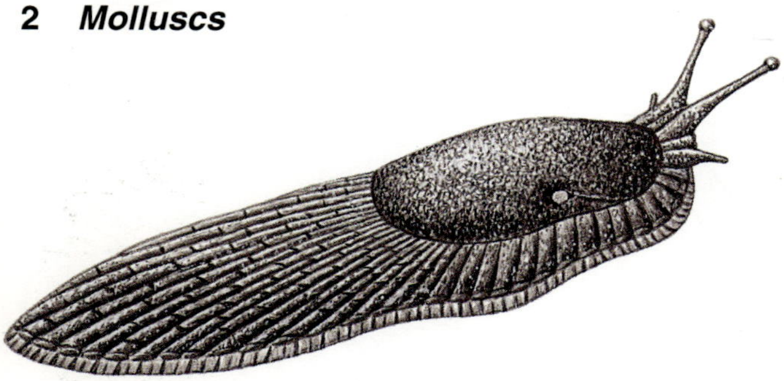

Slug
(1–10 cm)

Snail
(1–10 cm)

3 *Arthropods*

a Crustacea

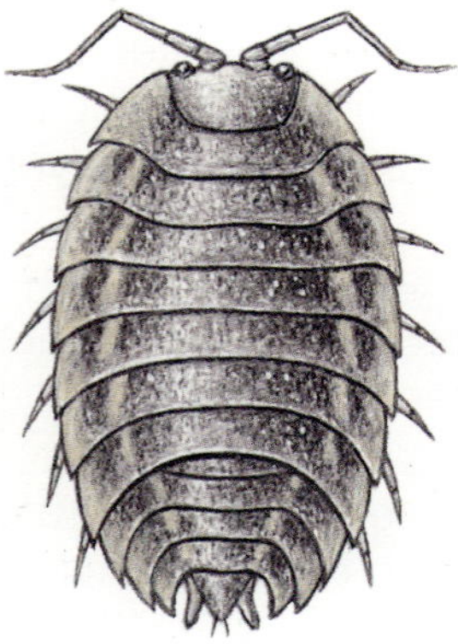

Woodlouse
(1 cm)

b Myriapods

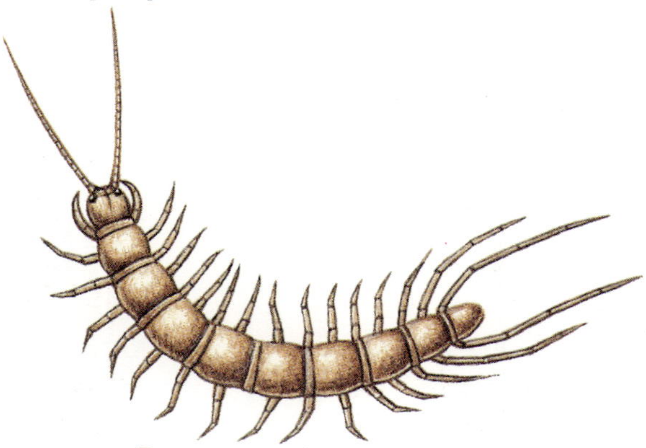

Brown centipede
(3–4 cm)

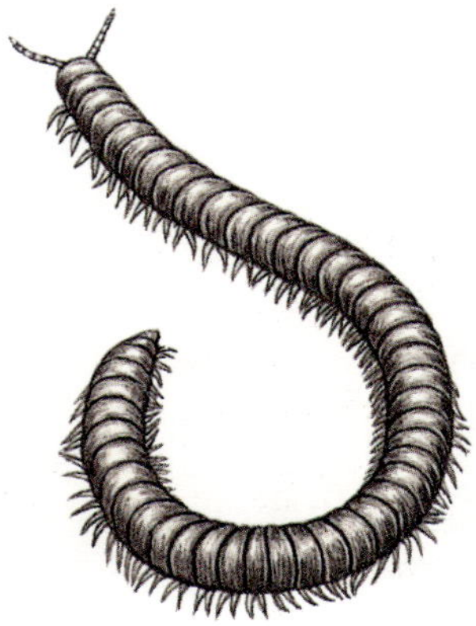

Snake millipede
(3 cm)

Arthropods continued.

c Arachnids

Garden spider
(Total length 1–2 cm)

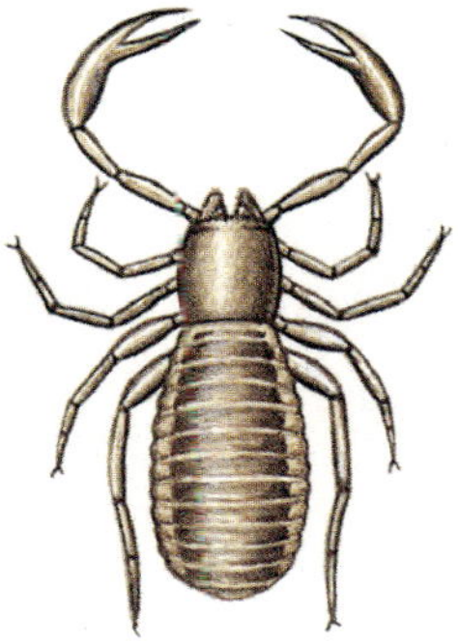

False scorpion
(0.5 cm)

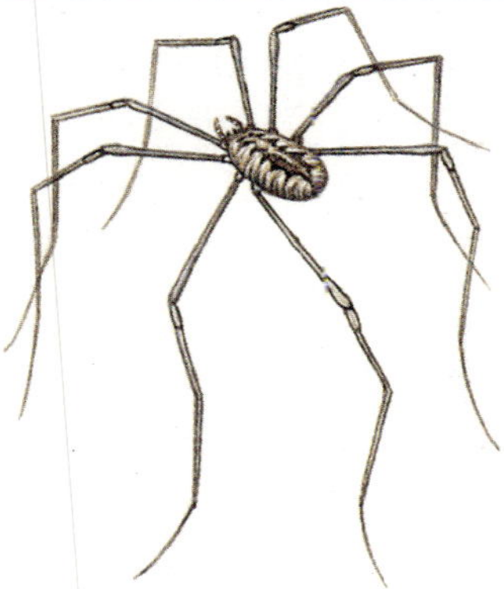

Harvestman
(1 cm)

Mite
(less than 1 mm)

d Insects

Insects have an unusual **life cycle**. They hatch from eggs as **larvae**. The larvae feed until they reach full size and then they form a **cocoon** around themselves. Later, the adult insect comes out of the cocoon.

Larvae

Rove beetle larva
(2 cm)

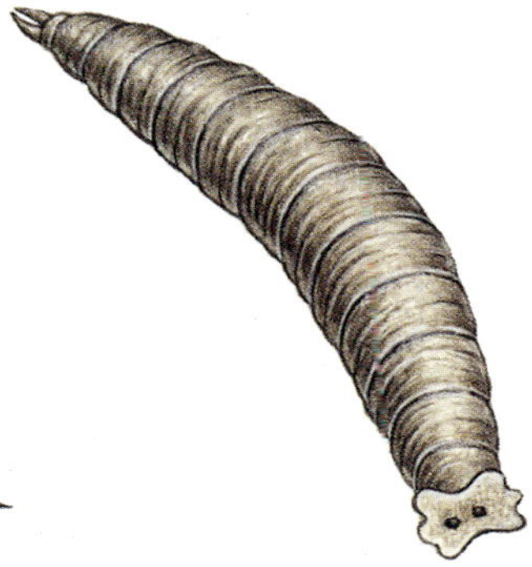

Leather jacket
(4 cm)

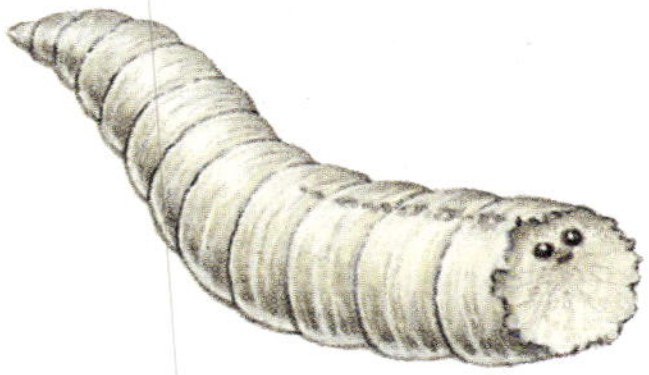

Fly maggot
(0.5 cm)

Weevil larva
(1 cm)

Adult insects

Rove beetle
(1–2 cm)

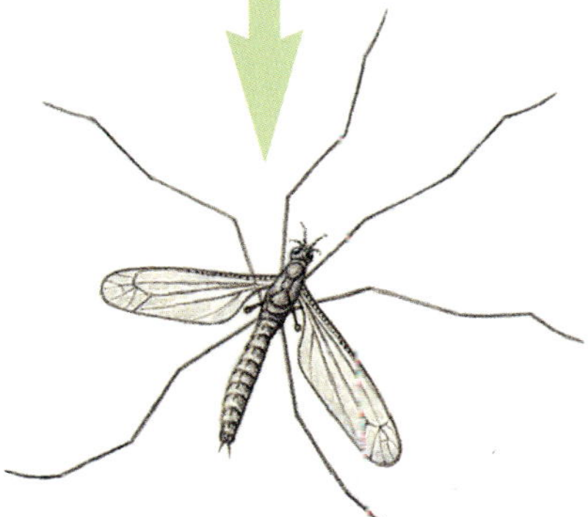

Crane fly
(2 cm)

Bluebottle
(1 cm)

Weevil
(1 cm)

Extension exercise 2A can be used now.

Sorting out Vertebrates

Vertebrates are animals which have a **backbone**. They vary in shape and size from mice to whales. All vertebrates can be put into one of five sets – **fish**, **amphibians**, **reptiles**, **birds** or **mammals**.

In this exercise you will use a series of questions (*a key*) to put each vertebrate in its correct set.

Mackerel

Cat

Q1 Copy this table.

Fish	Amphibians	Reptiles	Birds	Mammals

A Study these ten vertebrates. Use the key on page 11 to sort them into sets. ▼

Grass snake

Frog

Turtle

Great crested newt

Sparrow

Duck

Dolphin

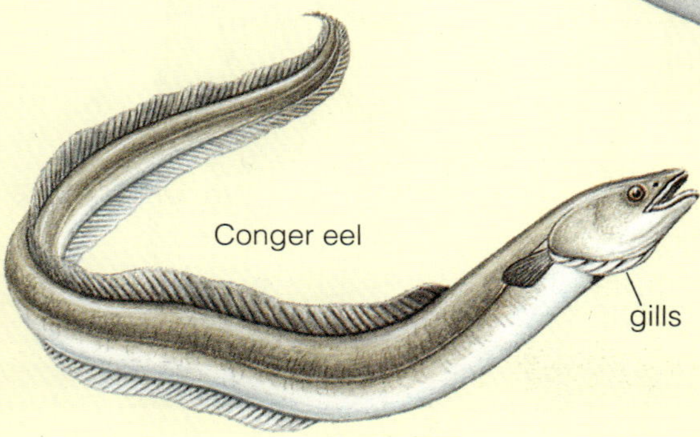

Conger eel

Useful information:
Some vertebrates (birds, mammals) can keep their body temperature constant even when their surroundings are very hot or very cold. They are said to be **warm-blooded**. They are warm to the touch. The other vertebrates (fish, amphibians, reptiles) are **cold-blooded** because their body temperature is always the same as their surroundings. Usually they are cold to the touch.

For more information on vertebrates see page 12.

A key to vertebrates

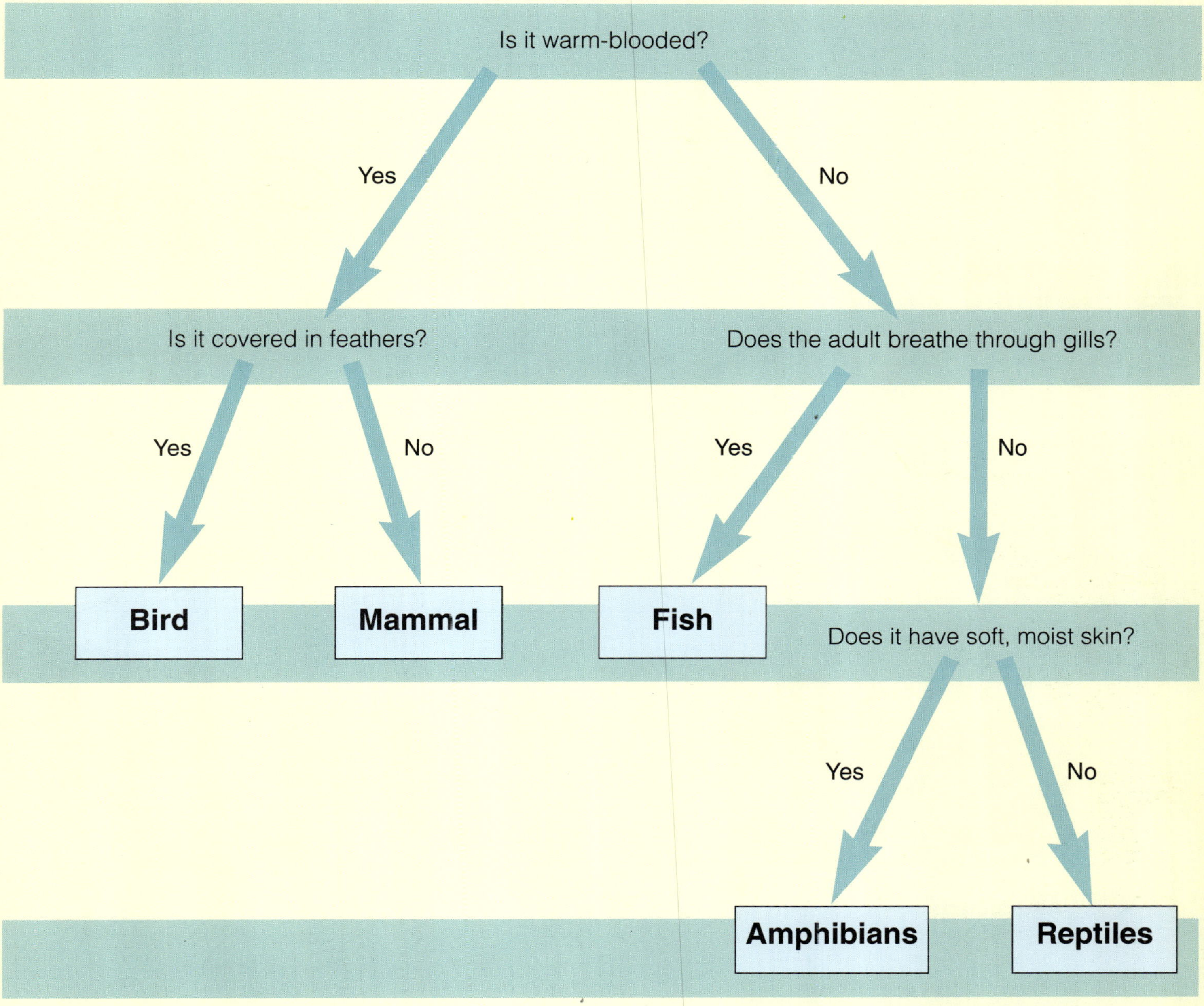

B Put each vertebrate in the correct column of your table. ▲

Q2 Name two fish.

Q3 Name two amphibians.

Q4 Name two reptiles.

Q5 Which sets breathe air?

Q6 Name a mammal that lives in the sea.

Q7 Name two birds that cannot fly.

Q8 How does a polar bear keep warm in the Arctic?

Q9 Why are there no reptiles living in cold regions like the Arctic?

Extension exercise 2B can be used now.

Vertebrates

All animals with backbones (vertebrates) can be put into one of five sets.

Vertebrates
- ☐ animals with backbones

Fish
- ☐ body covered in scales
- ☐ breathe with gills
- ☐ cold-blooded
- ☐ live in water
- ☐ move with fins

Reptiles
- ☐ dry scaly skin
- ☐ breathe with lungs
- ☐ cold-blooded
- ☐ lay eggs with leathery shells

Mammals
- ☐ hairy skin
- ☐ breathe with lungs
- ☐ warm-blooded
- ☐ feed young on milk

Amphibians
- ☐ moist skin with no scales
- ☐ adults breathe with lungs
- ☐ cold-blooded
- ☐ lay eggs covered in jelly
- ☐ live on land and in water

Birds
- ☐ skin covered in feathers
- ☐ breathe with lungs
- ☐ warm-blooded
- ☐ lay eggs covered in shell

Q1 Which set of vertebrates breathes with gills?

Q2 Which set of vertebrates has feathers?

Q3 Which sets of vertebrates are cold-blooded?

Q4 What does 'warm-blooded' mean?

Q5 Name two ways in which reptiles and amphibians are different.

Q6 During an expedition to the Solomon Islands, scientists discovered a strange animal living deep in the forest. It was furry, about 15 cm long and it had leathery wings. It felt warm when it was picked up. To which vertebrate set does it belong?

2 The balance of nature

Pond life

So far you have studied plants and animals living in the school grounds and in leaf litter. Both these places are called **habitats**. Another example of a habitat is a pond. In this experiment, you are going to investigate the living things in a pond.

Apparatus

- ☐ dropper ☐ beaker
- ☐ magnifying lens
- ☐ clipboard and paper
- ☐ fine mesh net ☐ plastic tray
- ☐ specimen tubes
- ☐ 10 metre measuring tape

In the laboratory:

- ☐ microscope
- ☐ cavity slide
- ☐ cover slip

Protect living things. Handle pond life with care.

A Look at your pond carefully. Draw a plan showing its size and shape. Mark on the plan such things as large stones, trees at the edge, plants above the water, plants below the water, or litter. ▼

B *Gently* sweep the net through the water. Put any animals you find into the plastic tray. Use the drawings on page 14 to identify them. **At the end of the lesson put the animals back in the pond.** ▼

C To catch very small animals, collect some water in the beaker. Use your dropper to suck up any tiny animals in the beaker. Put them into specimen tubes. ▼

Wash your hands thoroughly with soap and water after handling pond water.

D *Back in the laboratory*, put your very small (**microscopic**) animals onto cavity slides. Carefully cover your animal with a cover slip and study it using the low-power lens of the microscope. ▼

Q1 List all the animals you found in the pond.

Q2 What was the smallest animal you found?

Q3 What was the largest animal you found?

Q4 Why did you return the animals to the pond?

Q5 A water flea is a common pond animal. Watch it carefully and describe how it feeds.

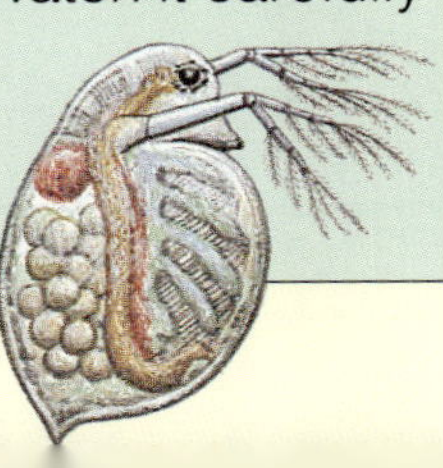

Water flea (magnified ×20)

Pond Life

Use these drawings to help you identify the living things you found in the pond. If in doubt, ask your teacher for advice.

They are not drawn to scale.

Animals

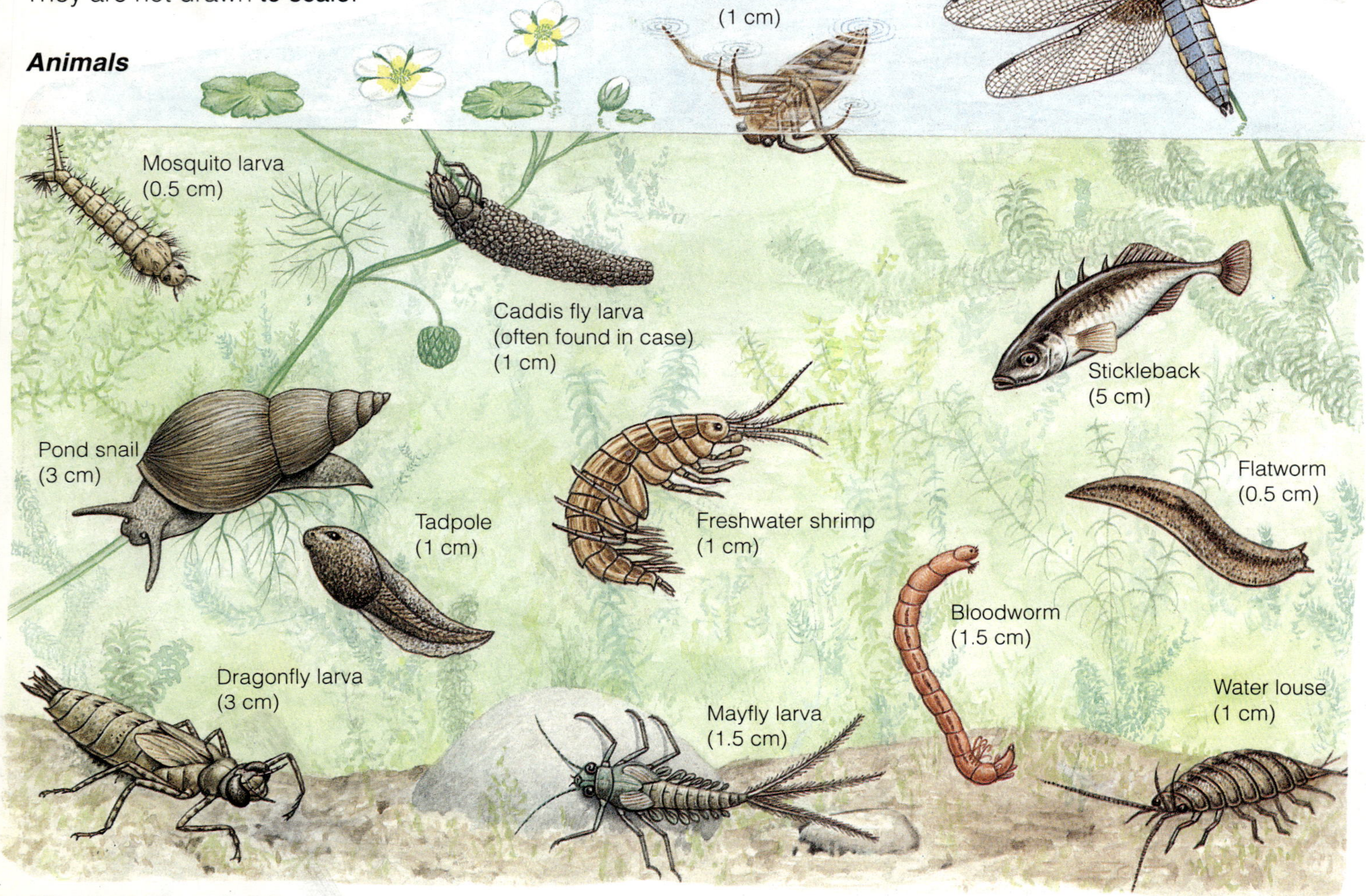

Microscopic animals

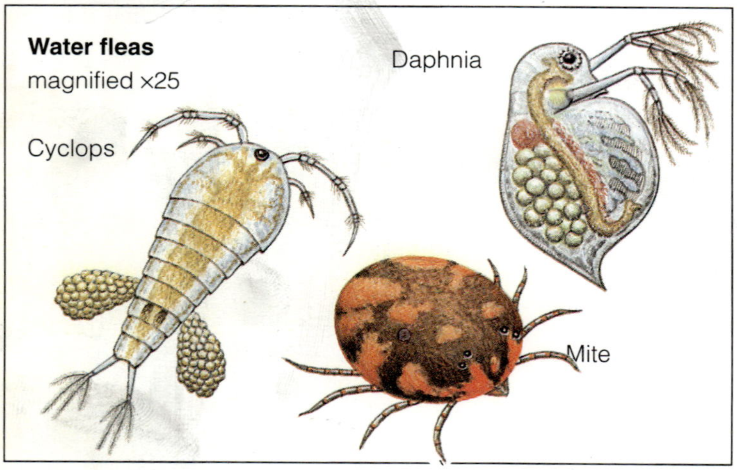

... and plants

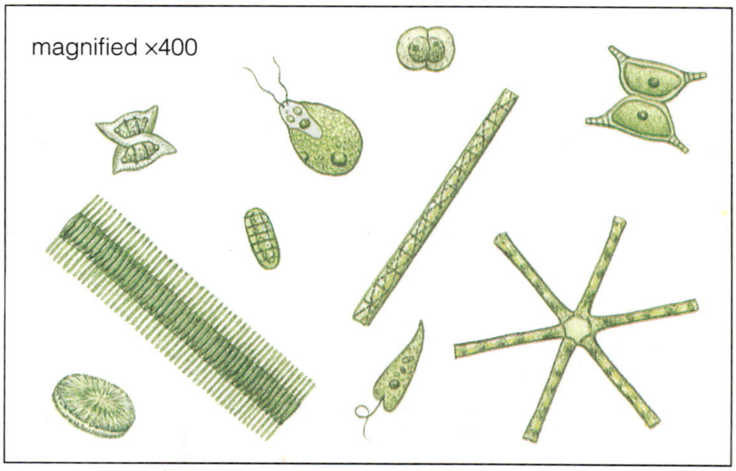

These tiny plants are eaten by many of the small animals.

What places do water fleas like best?

After a pond-dipping exercise, there was a discussion among the students about where the best place was to catch water fleas. Nadine said 'in clear water', Stephanie said 'under weeds' and Thomas said 'near stones and gravel'.

Design and *carry out* an experiment to find the water fleas' favourite place.

Apparatus provided

(You may not need all of it. Ask your teacher if you need extra apparatus.)

- ☐ plastic fish tank ☐ plastic tray
- ☐ pond water
- ☐ magnifying lens
- ☐ bench lamp ☐ sand ☐ gravel
- ☐ pond weed ☐ test tubes
- ☐ sample tubes
- ☐ 250 cm^3 beakers ☐ dropper
- ☐ water fleas

Protect living things. Handle the water fleas with care.

A In your group discuss how you can carry out the investigation. Write out your plan for the investigation. ▼

B Discuss your plan with your teacher. Make sure everyone in the group knows what to do. When your teacher has approved your plan, go ahead with your investigation. ▼

C Write a report of your investigation, explaining how you did the experiment and what results you got. ◀

Your report will need to answer these questions:

- ☐ How did you investigate the problem?
- ☐ What conditions do water fleas like best?
- ☐ How accurate do you think your results are?
- ☐ How could you improve your experiment?

Plants and photosynthesis

In our pond habitat, animals need plants to survive. The plants provide the animals with **food** and **oxygen**.

When light shines on the pond, the plants in it **photosynthesise**. They use sunlight energy to turn **carbon dioxide** and water into sugar. Oxygen is made as a **by-product**.

Carbon dioxide + water + sunlight energy ⟶ sugar + oxygen

Most of the sugar is changed to **starch**. It is stored in the plant's leaves. Plant-eating animals (**herbivores**) eat the leaves and use the starch for energy.

You can *prove* leaves make starch in the following experiment.

Apparatus

- ☐ ethanol ☐ boiling tube
- ☐ 400 cm^3 beaker
- ☐ Bunsen burner
- ☐ heatproof mat ☐ white tile
- ☐ Petri dish ☐ stop clock
- ☐ tweezers
- ☐ iodine solution and dropper
- ☐ matches and splints
- ☐ tripod and gauze
- ☐ plant A kept in light for two days ☐ plant B kept in dark for two days ☐ eye protection

Safety Warning
Wear eye protection when heating liquids.

Ethanol is flammable. Keep it away from naked flames.

A Choose two similar leaves, one from plant A and one from plant B. ▲

B **Wear eye protection**. Take the leaf from plant A and put it into a beaker of boiling water for a few seconds. The boiling water softens the leaf. ▶

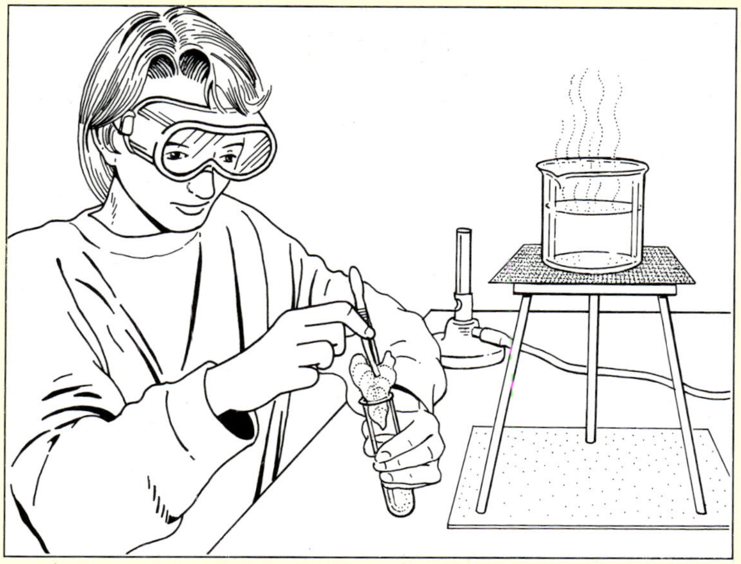

C **Turn off the Bunsen burner. Do not use the ethanol until all the flames are out.** Take the leaf from the boiling water using tweezers and put it into a boiling tube half-full of ethanol.

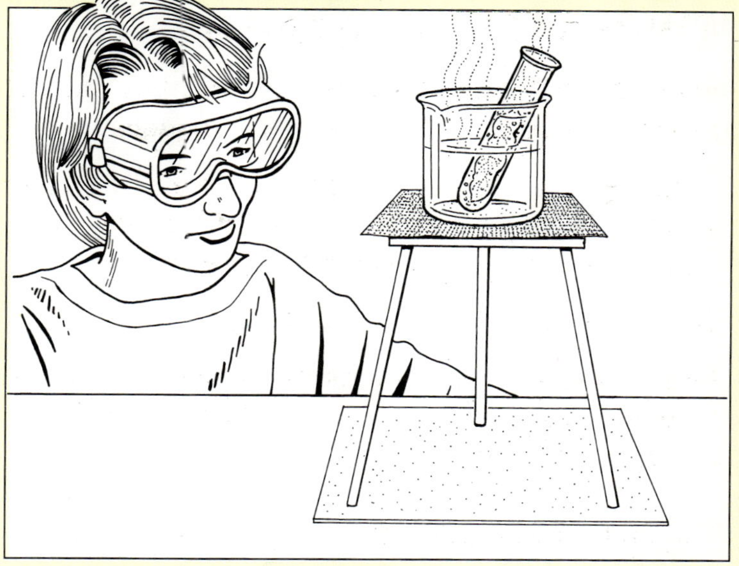

D Put the boiling tube containing the leaf and ethanol into the beaker of hot water. Leave it there for two minutes until most of the green colour (**chlorophyll**) has come out of the leaf.

E Take the leaf from the ethanol using tweezers. Put it into a Petri dish of cold water to soften it.

F Lay the leaf flat on a white tile and cover it with drops of **iodine** solution. Starch turns iodine dark blue or black.

G Repeat **B** – **F** for the leaf from plant B.

Q1 Which leaf contains starch?

Q2 What does a leaf need to make starch?

Q3 Which gas is made during photosynthesis?

Q4 How do animals use this gas?

Q5 In the school entrance hall, there is a large fish tank containing 20 small tropical fish. The air bubbler has broken and the school cannot afford a new one.

Ms Smart, a science teacher, says that if there are plenty of plants in the tank and it is well lit, you don't need a bubbler. The headteacher is not happy with this idea.

Can you explain to the headteacher what Ms Smart means?

Extension exercise 3 can be used now.

Who's eating who?

All life in the pond depends on plants making food by photosynthesis. This food is used in one way or another by all the animals in the pond.

▲ Very small plants called microscopic algae photosynthesise and make starch. Small animals (*e.g.* insect larvae) feed on these plants. Small fish (*e.g.* sticklebacks) feed on the larvae. Then the small fish are eaten by bigger fish (*e.g.* perch). This process is called a **food chain**.

▶ In a food chain, the plants and animals are given different names to describe how they feed.

▼ Often, food chains are complicated by other animals sharing the same food. This is called a **food web**. All the plants and animals in your pond depend on each other.

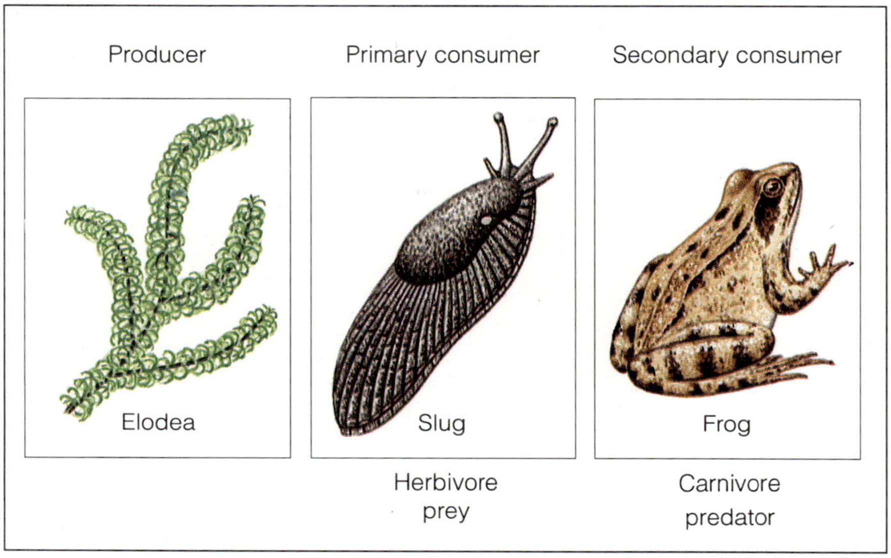

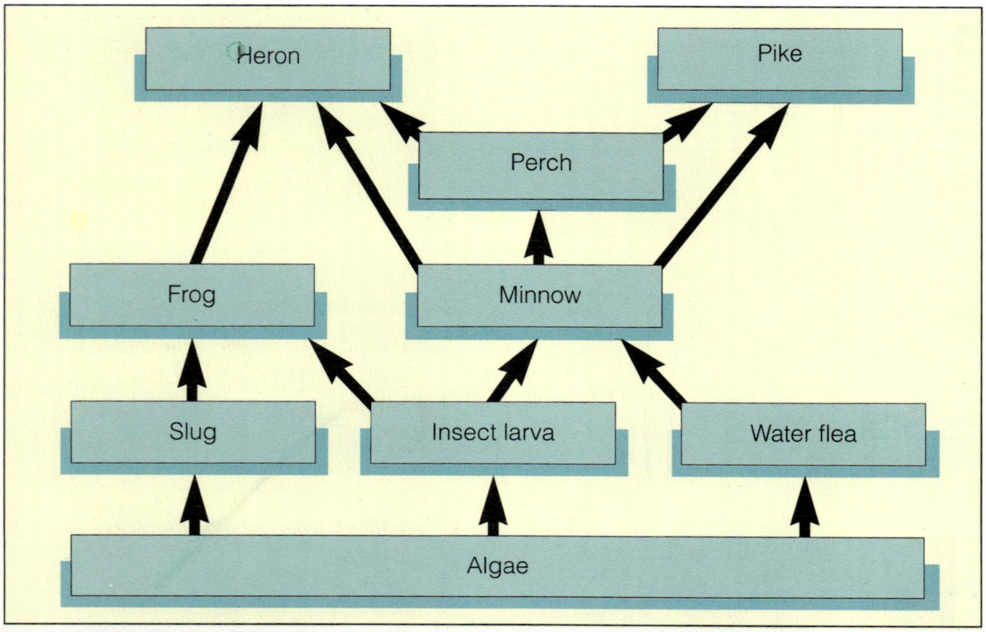

Q1 What is photosynthesis?

Q2 Name two herbivores.

Q3 Name two carnivores.

Q4 In nature a food web is balanced. There are enough herbivores to feed the carnivores. What would happen if more plants were put into a pond?

Q5 A fishing club has decided to put five young perch into a local pond. How will this affect the balance of the food web?

Extension exercises 4 and 5 can be used now.

Where the energy goes

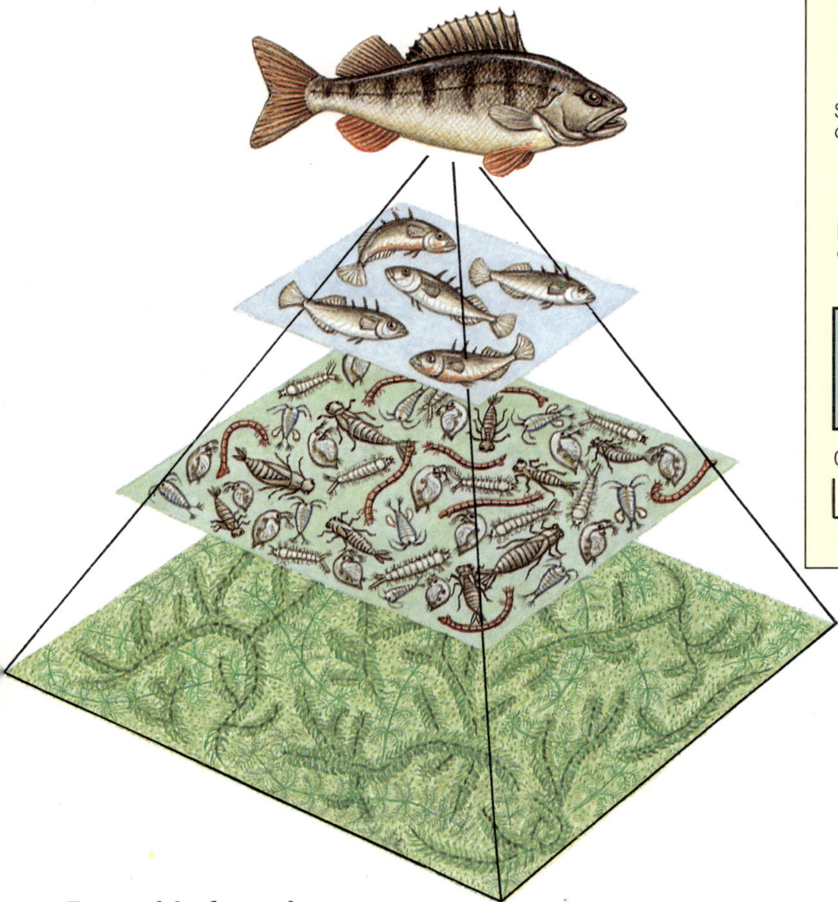

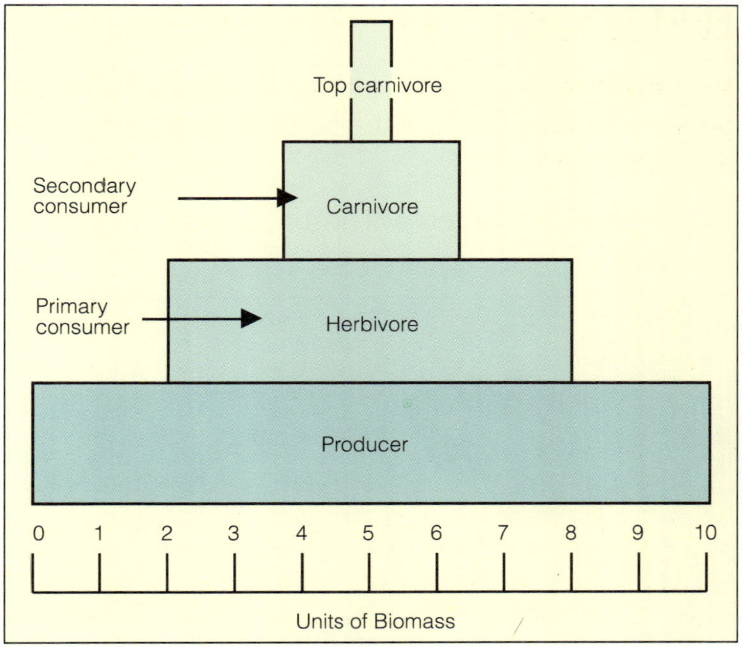

Pyramid of biomass

▲ Another way of studying a food chain is to weigh all the plants or animals at each stage. The weight of a plant or animal is called its **biomass**. In many food chains the pattern of results is like the diagram above.

Pyramid of numbers

▲ When we count the number of plants and animals in a food chain, we find that there are a great many needed at the beginning of the chain to feed a few at the end. This is called a **pyramid of numbers**.

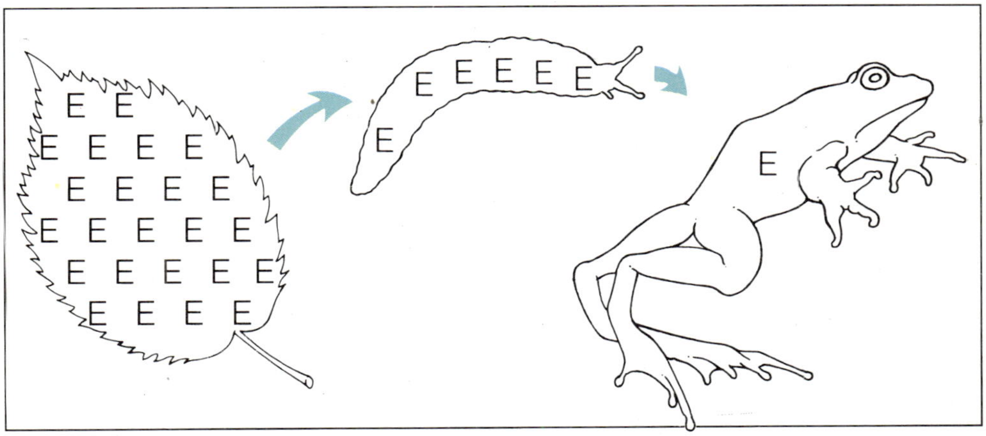

Energy transfer

▲ In this food chain, the plant traps the sun's **energy**. The slug eats the plant. It uses some of the energy from the plant and stores the rest. When the frog eats the slug, it gets some of the energy stored in the slug. At each of these steps some energy is wasted and lost from the food chain.

Q1 What living things are at the base of the pyramid of numbers?

Q2 What living things are at the top of the pyramid of numbers?

Q3 What is meant by the biomass of an animal?

Q4 Look at this food chain:

chicken
↗ ↘
rice → humans

We can get energy by eating the rice. Or we can feed the rice to chickens and then eat the chickens. Which path is the most *efficient* way to get energy?

Q5 Why does meat cost so much?

Hedgerows

A **hedgerow** is a line of hedges and bushes which grow along the side of a field or road. A hedgerow is home to many plants and animals.

In this experiment you will find out which animals live in the hedgerow.

Apparatus

- ☐ fine mesh sweep net
- ☐ trowel ☐ card lids ☐ pooters
- ☐ small soft paintbrush
- ☐ specimen tubes
- ☐ yoghurt pots ☐ newspaper
- ☐ clipboard and paper

Protect living things. Handle the animals with care.

A Quietly watch your hedgerow. Make a note of any animals you see. ▲

B Walk slowly along the side of the hedgerow. Use the sweep net to catch any invertebrates you disturb. Carefully brush them into your specimen tubes. ▶

C Set **pitfall traps** at the bottom of the hedgerow. Leave the traps overnight. Carefully brush any invertebrates you catch into your specimen tubes. ▲

D Use a pooter to collect any small invertebrates you can find in the hedgerow. ▲

E Place a sheet of newspaper below one side of a bush. *Gently* shake or tap the bush. Collect any invertebrates that fall from the bush in your specimen tubes. ▲

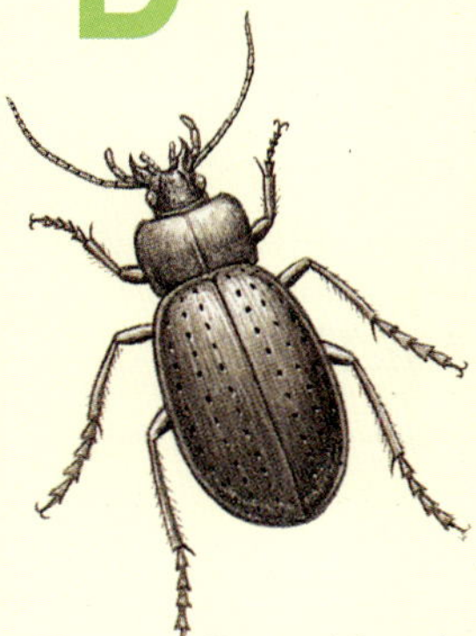

Ground beetle (2.5 cm)

Butterfly (2 cm)

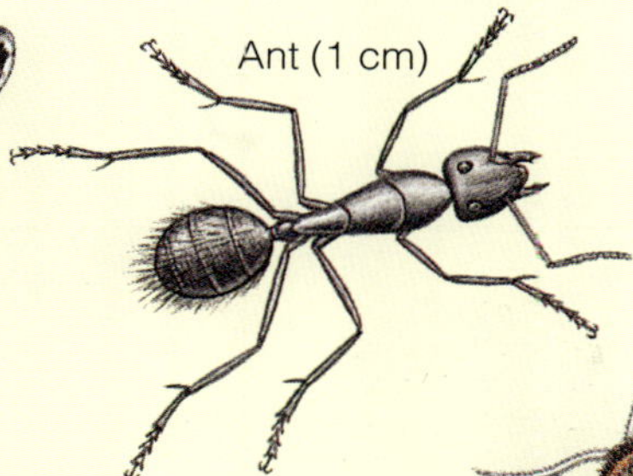

Ant (1 cm)

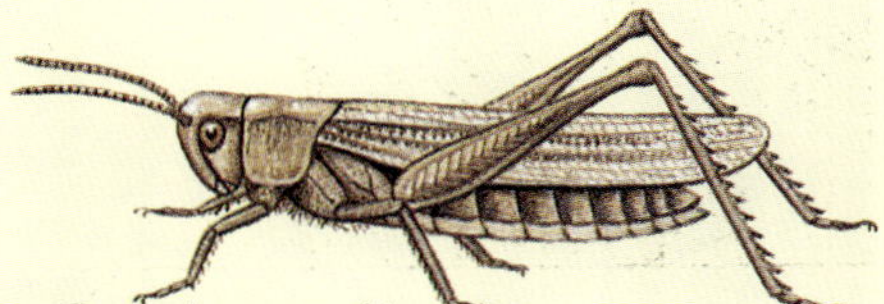

Grasshopper (1 cm)

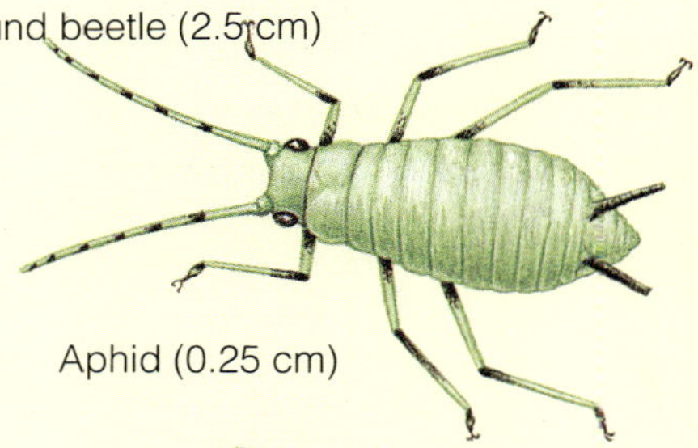

Aphid (0.25 cm)

Ladybird (0.5 cm)

Moth (2 cm)

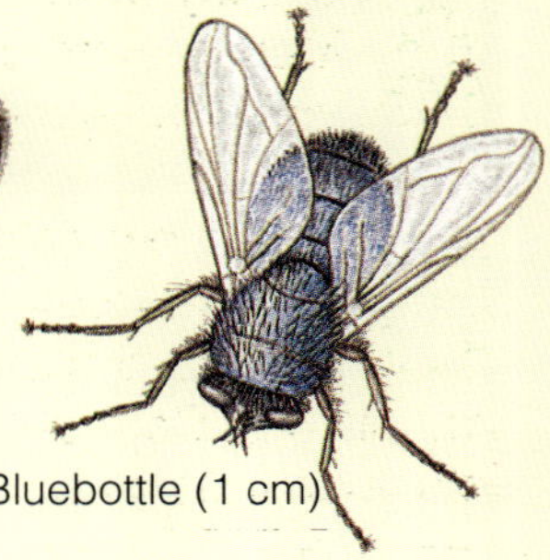

Bluebottle (1 cm)

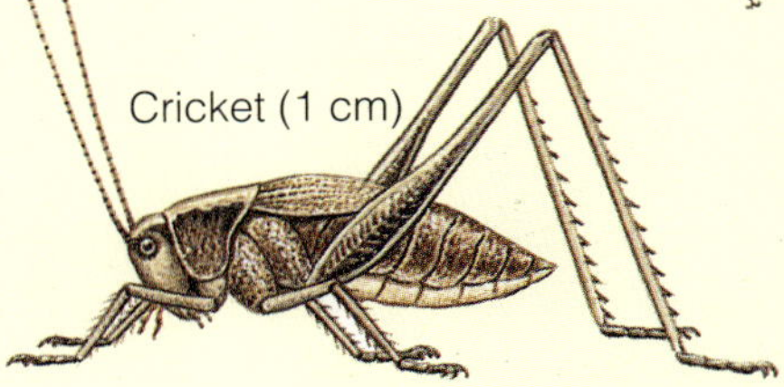

Cricket (1 cm)

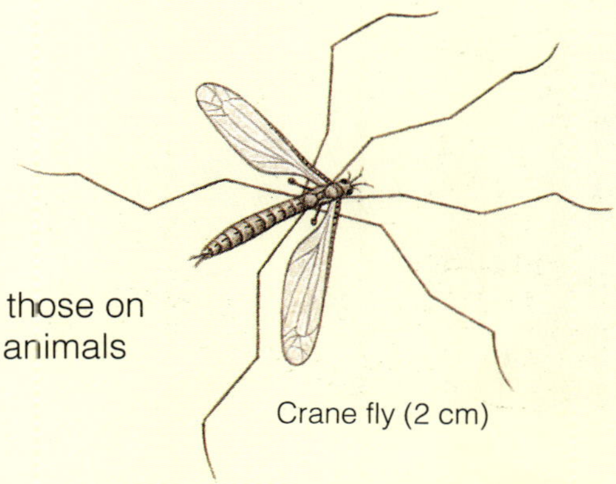

Crane fly (2 cm)

F Use the drawings above and those on pages 8 and 9 to identify the animals you have found. ▲

Q1 List all the animals you have found.

Q2 Name two decomposers living in your hedgerow.

Q3 Draw a food web for your hedgerow. Include in it as many of your animals as possible.

Disappearing hedgerows

▲ In your survey of a hedgerow, you saw some of the animals which live there. Hedges are very popular with insects, birds and small mammals for feeding and **breeding**. They are not so popular with **farmers**.

▲ Hedges divide farmland into small fields. Taking away the hedges turns three or four small fields into one big one. Large fields are easier to work with machines and there is less wasted land.

▲ Since 1947, a total of 180 000 km of hedgerow (long enough to go round the Earth four times) have been ripped up. In some parts of this country, as many as five small farms have been turned into one single field.

Fortunately, things are slowly improving. Fewer hedges are being ripped up. More farmers are aware of the problem of birds and butterflies disappearing from farmland.

▶ Another reason is that farmers have produced too much food. We have 'mountains' of stored butter, meat and **grain**. We don't need so many giant fields.

Q1 What do birds use hedgerows for?

Q2 Why do farmers rip up hedgerows?

Q3 How many kilometres of hedgerow have been destroyed since 1947?

Q4 What is meant by a grain mountain?

Q5 This country grows more food than it eats. All the extra meat, butter and grain has to be stored in warehouses. This is expensive and wasteful, and the food mountains get bigger every year. What can we do about it?

Plant nutrients

▶ Plants get most of their food by photosynthesis. They also need small amounts of **mineral salts** to grow well. Two important salts are **nitrates** and **phosphates**.

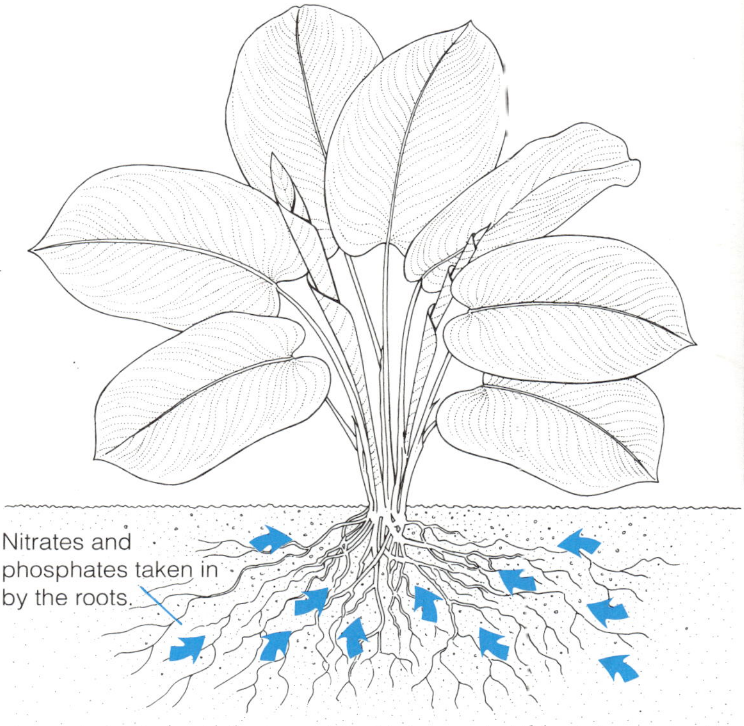

▲ Under the hedgerow the soil is rich in mineral salts from the rotting bodies of dead plants and animals. **Decomposers** (*e.g.* **bacteria**, **fungi**, slugs, beetles) feed on this dead material and break it into small pieces. The rotting pieces (**humus**) are rich in salts. These salts feed the young plants.

▲ A gardener uses the same idea to make humus. Waste from the kitchen and garden is collected in a **compost** heap. Here decomposers get to work. They rot the waste and release the mineral salts. Finally, the compost is mixed into the soil to feed growing plants.

▲ Most farmers cannot make enough compost to feed their large crops. Usually they use **fertilisers** made in **chemical** factories. They spread these chemical salts over the field to feed the crops.

Q1 Name two important salts needed by a plant for healthy growth.

Q2 Name three decomposers.

Q3 What do decomposers feed on?

Q4 What is compost?

Q5 Where do farmers normally get their fertiliser from?

Q6 A local health food shop is advertising **organically grown** vegetables. Find out what this means. Explain why some people prefer them to normal vegetables.

Extension exercise 6 can be used now.

Investigating plant nutrients

Plants need a mixture of mineral salts to grow well. In this experiment, you will see what happens when some of these mineral salts are missing.

Apparatus

- ☐ nine flasks ☐ cotton wool
- ☐ tray of seedlings (barley if available)
- ☐ plant culture tablets

Your teacher will:

A Label the nine flasks as shown. Put the correct tablet into flasks 1–8. Add distilled water to all nine flasks. Shake flasks 1–8 to **dissolve** the tablets. ▼

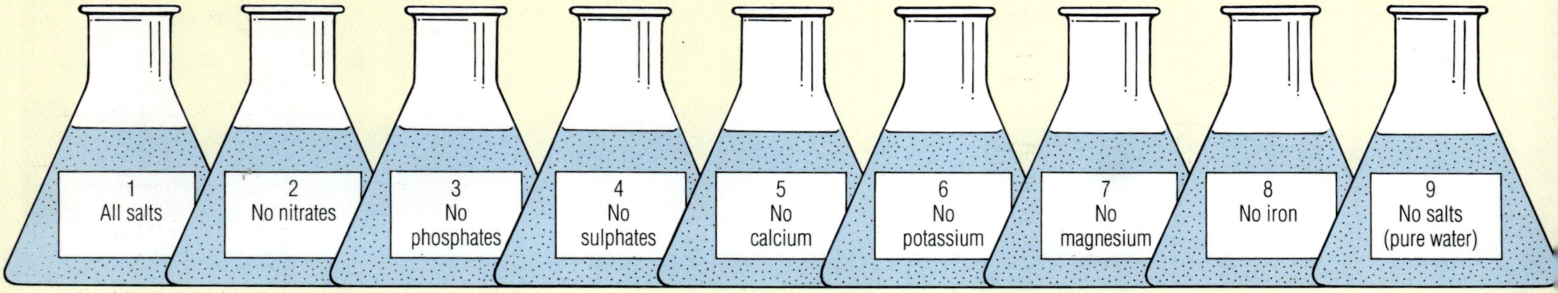

B Choose nine healthy seedlings of about the same size. Put them into the necks of the flasks. Hold them in place with cotton wool. ▼

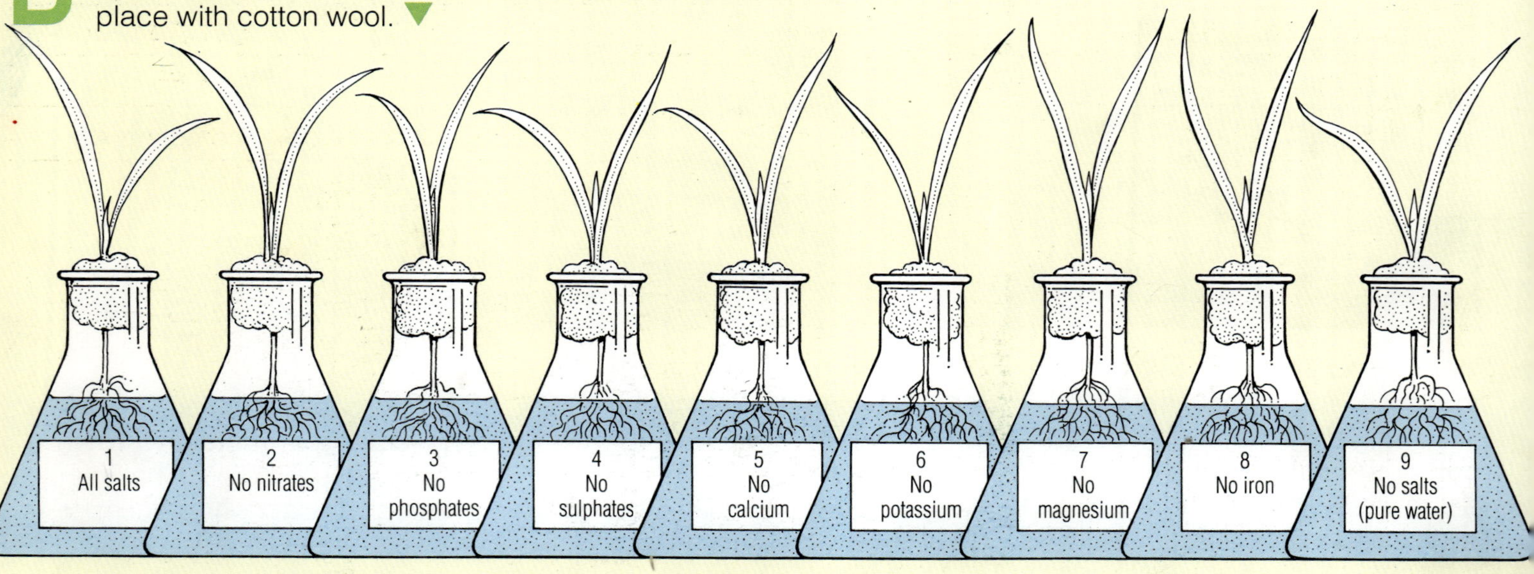

C Leave the seedlings in a warm, light place for several weeks.

D After several weeks, copy and complete the table below: ▼

Flask number	Missing salt	Appearance of seedling
1	all salts present	
2	nitrates	
3	phosphates	
4	sulphates	
5	calcium	
6	potassium	
7	magnesium	
8	iron	
9	no salts (pure water)	

Q1 In which flasks did the seedlings grow well?

Q2 In which flasks did the seedlings grow badly?

Q3 Which salts are needed by plants for healthy growth?

Garden centres sell packets of fertilisers containing mineral salts which gardeners can use in their garden. Unfortunately the instructions are sometimes very complicated. Study this table and then answer the questions.

Fertiliser	Type of nutrient	Amount (g per m^2)	Cost of 1 kg packet
Hoof & horn	nitrate	100	£3.20
Sulphate of ammonia	nitrate	50	£5.50
Bone meal	phosphate	100	£2.85
Super phosphate	phosphate	100	£3.30

Q1 How much hoof and horn fertiliser would you use on one square metre of garden?

Q2 How much bone meal would you use on 10 m^2 of garden?

Q3 How much will it cost to fertilise a 10 m^2 garden using:
a hoof and horn?
b sulphate of ammonia?

Garden fertilisers

Carol is interested in gardening. She looks after the garden at home. Her lawn is looking worn and she has decided to use a fertiliser on it.

Carol is a bit mean. The instructions with the fertiliser tell you how much to spread on every square metre. She thinks she can get away with using a lot less.

Design an experiment to find out how much fertiliser gives the best results on grass.

Apparatus provided

(You may not need all of it. Ask your teacher if you need extra apparatus.)

- ☐ lawn fertiliser
- ☐ electronic balance ☐ string
- ☐ watering can ☐ metre rules
- ☐ tent pegs ☐ spatula
- ☐ hammer

A Discuss your plan in your group. Think about questions such as
'Is it fair?'
'Is it accurate?' ▲

B Check your plan with your teacher. If there are no problems go ahead with your investigation. ▲

C Write a report about your experiment. Your report should answer questions such as: ▶

- ☐ How many test areas did you use?
- ☐ How big was each test area?
- ☐ How much fertiliser did you put on each test area?
- ☐ How long did you leave the test area?
- ☐ How did you check the growth of the grass?
- ☐ What amount of fertiliser is best for 1 square metre of grass?

Extension exercise 7 can be used now.

3 Threats to wildlife

Chemical fertilisers

Chemical fertilisers, rich in **nitrates** and **phosphates** are used on most farms in this country. They are used to produce a bigger crop. Chemical fertilisers are cheap and easy to use but they do cause problems.

▼ Many of the chemicals in fertilisers (especially nitrates) dissolve in water. When it rains, some of the fertiliser is washed out of the soil and into the streams and rivers.

▼ The dissolved fertilisers can affect life in the river. They make river plants grow too quickly. Soon there are so many plants that the river becomes **choked**. The plants die and as they rot they use up the oxygen needed by the fish.

▲ Every year Water Authorities spend thousands of pounds removing these choking water plants. It is a problem which will not be solved until farmers stop using so much chemical fertiliser.

▼ Chemical fertilisers are rich in mineral salts but they do not contain humus. Without humus, soil becomes dry and dusty. High winds can then blow the soil away.

Q1 What is the major advantage of chemical fertilisers?

Q2 What happens to fertilisers like nitrates when rain falls on a field?

Q3 What effect does the fertiliser in the river water have on the river plants?

Q4 What do Water Authorities do about the rivers choked with plants?

Q5 What is humus?

Q6 What happens to a soil without humus?

Q7 Many Water Authorities take drinking water from rivers. What effect can the use of large amounts of fertiliser have on our drinking water?

Nitrate pollution

When nitrate salts from fertilisers are washed from fields into rivers, they can cause a great deal of harm. The nitrates in the river water feed the water weeds. The weeds grow thickly and choke the river.

▶ As the weeds die, bacteria grow on them. The bacteria use up the oxygen in the water leaving little for the fish. Without oxygen the fish start to die.

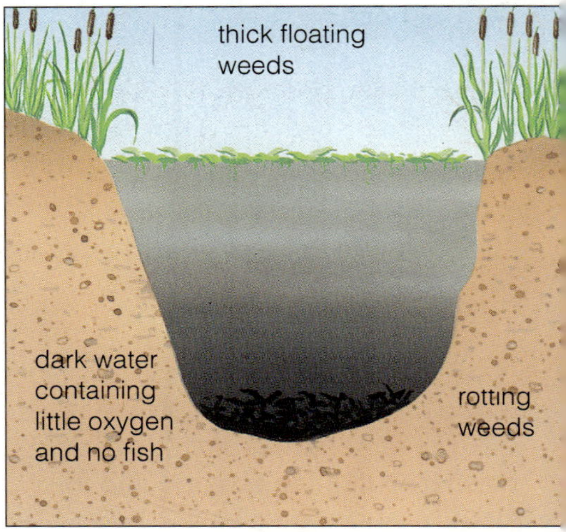

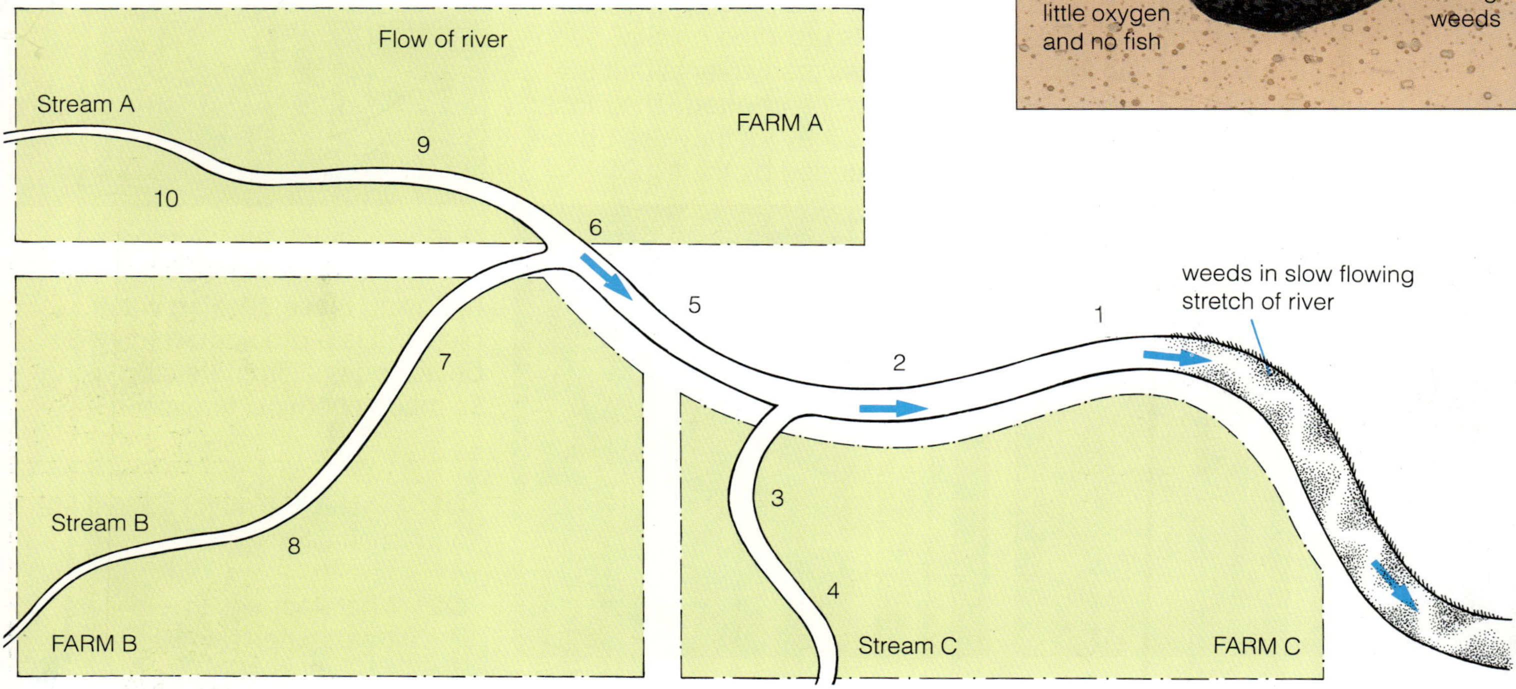

Solving a problem

▲ A Water Authority has a serious problem in one of its small rivers. The lower reaches of this slow-running river are choked with weeds. Each year the weeds are cut back but they soon grow again. The Water Authority has decided to do something about it.

▶ Scientists took water samples from ten places along the river. They **analysed** the water samples for fertiliser nitrates. The results are shown in this table.

Position	Nitrate level (parts per million)
1	85
2	90
3	5
4	5
5	100
6	110
7	10
8	10
9	250
10	210

Q1 Where do you think the nitrate pollution is coming from?

Q2 This river is a slow-flowing river. Would there be the same problems if it was a fast-flowing river? Explain your answer.

Q3 What can be done to solve the problem?

Extension exercise 8 can be used now.

Pesticides

As well as using chemical fertilisers to feed the crop, farmers also use chemicals called **pesticides** to kill **pests**. These pests damage the crop.

▼ There are many sorts of pesticide. **Insecticides** are used to kill insects feeding on the crop. **Herbicides** are used to kill weeds growing in the crop. **Fungicides** are used to stop **moulds** and fungi growing on the crop.

▼ Sadly, the pesticides affect much of the wildlife on the farm. Many of the animals in the hedgerow are poisoned by pesticides. Some pesticides are still **poisonous** many months after being sprayed.

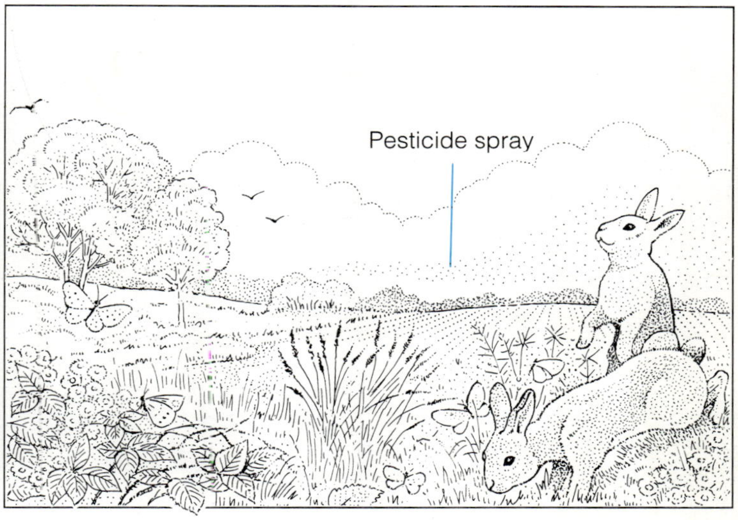

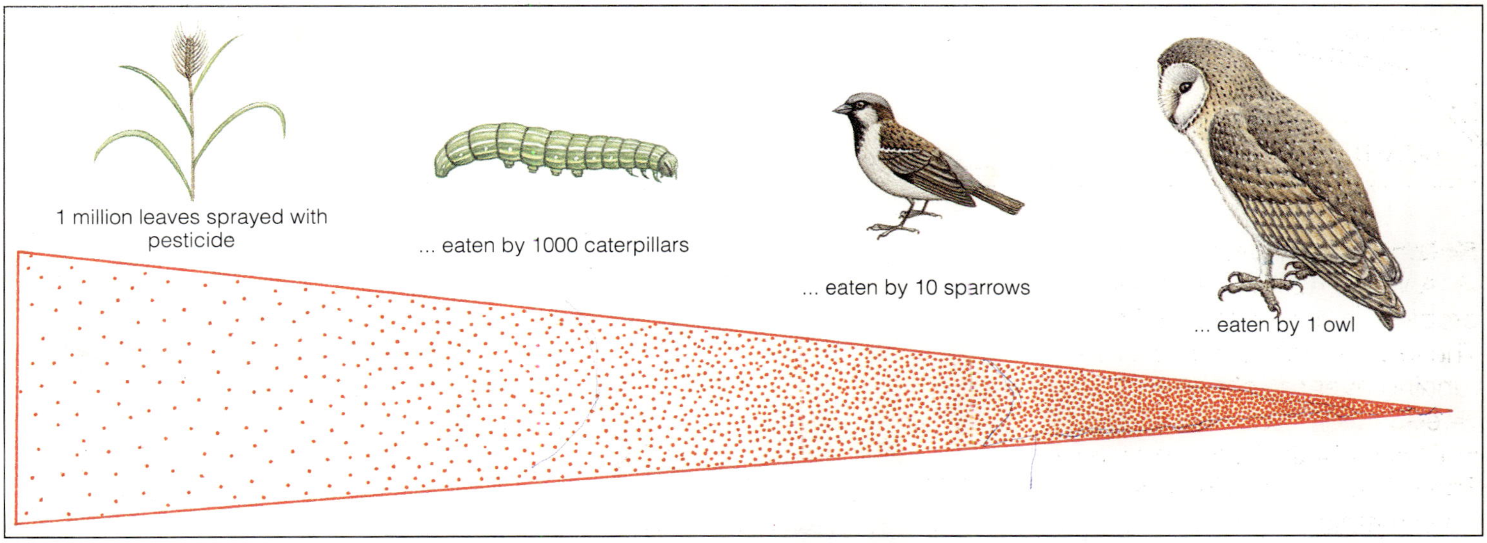

Pesticides and food chains

▲ When a crop of barley is sprayed with pesticide, the effect can be far reaching.

As the pesticide passes down the food chain it becomes more **concentrated**.

By the time it reaches the owl, it is strong enough to kill. Since 1968, 50% of the barn owl population has been wiped out.

Q1 What is a herbicide?

Q2 What is a fungicide?

Q3 Why do farmers use pesticides?

Q4 How can a large bird like an owl be killed by a spray used to kill tiny insects?

Forests and trees

We use plants and animals for many of the materials we need. They give us food, clothing and building materials.

▲ A thousand years ago, this country was covered in forests. The forest was home to thousands of plants and animals. People used the trees for firewood and cleared the forest to make farmland. Today, most of the forests have gone and so have the plants and animals that shared this habitat.

▲ Every week, especially in South America, thousands of **hectares** of **rainforest** are destroyed for timber and farmland. Every week, species of plants and animals disappear forever.

▲ Trees provide people with timber, but they also make a life-giving gas – **oxygen**. Animals need oxygen to stay alive. If we cut down all the trees we will also destroy animal life.

▲ What can be done to save the rainforests? One solution is to make better use of the materials we get from them. Most trees are used to make paper. The paper is used once and then thrown away. But it can be **recycled** and used again.
Another solution is to **replant** forests with young trees. These young trees will be the forests for your children and your grandchildren.

Q1 Name three things plants and animals give us.

Q2 When was this country covered in forests?

Q3 What happens to the forest animals when the trees are cut down?

Q4 Which gas is made by plants?

Q5 Why do animals need plants?

Q6 What can be done to stop the **destruction** of the South American rainforests?

4 Conserving wildlife

Endangered species

People are changing the countryside to meet their needs:

Hedges and woods are pulled up to make farmland
Marshes are drained to make farmland
Forests are cut down for timber
Habitats are poisoned by pesticides
Waterways are polluted by fertilisers

Many species of plants and animals are at risk of dying out. The plants and animals are said to be **endangered**. Once a species dies out, it is lost forever.

Great Crested Newt

Mouse-eared bat

Nightingale

Otter

Barn Owl

Red-backed Shrike

Corncockle

Harvest Mouse

Q1 Name three endangered species.

Q2 What does 'endangered' mean?

Q3 Are the chances of **survival** for animals like the mouse-eared bat and the red-backed shrike good? Explain your answer.

Q4 Design and make a poster to make people aware that our wildlife is dying out. Make it eye-catching with a simple message.

Saving our plants and animals

Around the world some 25 000 plant species and over 1 000 animal species (not including insects) are threatened with **extinction**. Each day at least one species is lost forever.

▶ Since 1949, ten **National Parks** have been set up in this country to protect our countryside. These National Parks are large areas of beautiful and wild countryside with strict rules to control industry and farmers. People are encouraged to enjoy the countryside but are asked to respect it.

'Take nothing but photographs, leave nothing but footprints'

▲ **Nature reserves** have been set up round the country to provide safe areas for endangered species. Visitors can see the animals and plants in their natural surroundings without harming them.

▼ In 1981, the **Wildlife and Countryside Act** became law. It protects plant and animal life. The law makes it a crime to disturb or harm animals such as the otter, badger and red squirrel. Wild birds, their nests and eggs are protected, as well as many rare plants.

Q1 How many plant species are threatened with extinction throughout the world?

Q2 Where is your nearest National Park?

Q3 When was the Wildlife and Countryside Act made law?

Q4 Is it legal to collect birds' eggs?

Q5 Describe a visit to a local Nature Reserve. Draw some of the plants or animals you saw there. Explain what the reserve is doing to protect wildlife.